THE FUNCTION OF PROTECTION & INDEMNITY MARINE INSURANCE IN RELATION TO SHIP OWNER´S LIABILITY FOR CARGO CLAIMS

Framing the Legal Context

Joseph Tshilomb JK MSc & LLM

authorHOUSE®

AuthorHouse™ UK
1663 Liberty Drive
Bloomington, IN 47403 USA
www.authorhouse.co.uk
Phone: 0800.197.4150

Published by AuthorHouse   03/02/2016

ISBN: 978-1-5246-2882-6 (sc)
ISBN: 978-1-5246-2883-3 (hc)
ISBN: 978-1-5246-2884-0 (e)

Print information available on the last page.

This book is printed on acid-free paper.

CONTENTS

ACKNOWLEDGEMENT

The author would like to express his gratitude to individuals and organizations who assisted him with moral support and useful advices while working on the project. Without their valuable feedbacks, it would have been more difficult to complete the work.

In particular, he is grateful to AuthorHouse® for its wide-ranging support from the kind invitation to write a book on the subject to the valuable publishing resources which made possible the achievement of this literary endeavour. The encouragement and support offered by Debbie from Edinburgh have enhanced his integration in the British society. For her dedication to this work the author would like to express heartfelt gratitude!

The author wish particularly to thank the Degen Family in Stockholm and Family Hammar in Köping through this publication!

As the list of names for those who contributed to the project is not exhausted, the author extends his sincere appreciation to all those who have not been mentioned.

PREWORD

"Great dreams need more time and careful considerations for their successful achievement; it is advisable to proceed gradually until the desired building is raised and decorated."

These words were my advice to the author of the book in need of quality time!

I was well aware that Mr Joseph Tshilomb was researching in the complex area of International Trade Law.

He is a Swedish citizen and native from Elisabethville (L'shi). I have known Mr Tshilomb during many years and have had many encounters with him talking about ethical matters concerning global and environmental aspects. He has shown a strong spirit of cooperation and motivation to achieve his project on shipping, maritime law and international trade.

I would like to emphasize the author's kind and generous character. He is a cultivated gentleman and scholar. His mind is strong and stable; he argues in a tolerant and convincing way. He feels at home in international social surroundings. He is able to change easily to different languages depending on the persons participating in the conversation.

With all these matters in mind, I think I have good and many reasons to recommend the result of his research.

+ Erwin Bischofberger

PhD (University of Frankfurt)

Professor emeritus

Karolinska Institute

Stockholm, October 2011

PREFACE

In the early days of shipping and international maritime trade safe arrivals of ships were a real challenge. Ship owners and cargo owners were gathering together in coffee houses in London in the hope to hear good news about ships as much more casualties occurred at sea at that time. The concept of marine insurance emerged from these uncertainties related to the safe arrival of the ships carrying cargoes.

Ship owners needed to assure a safe voyage and provide the cargo owners with a conditional loan covering a substantial part of the goods carried on the ship. In the case of the safe arrival of the ship, the cargo owner will fully compensate the ship owner and pay back the conditional loan received at the loading of the cargo on the ship. Nowadays, the shipping industry has witnessed major technical developments.

On one hand, the industry has experienced real improvement in safety standards brought on by better port state control systems, higher standards of classification and the advent of the

International Safety Management Code. The increase in ship size and the significant progress in containerization have led to further improvements in the industry.

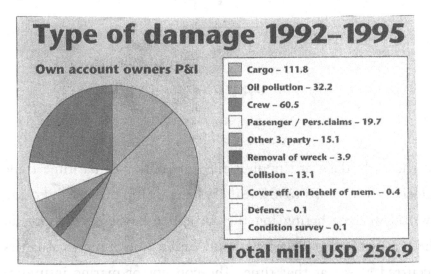

Hence, the number of casualties in shipping has generally gone down.

On the other hand, ship owners have seen their insurance costs rise, particularly on the liability side in this highly competitive international business requiring financial stability.

Ship owner's liability for cargo claims has been increasing both in number and in cost. In order to make it easier for a ship owner to operate in international shipping the protection and indemnity insurance appeared around 1870 as mutual marine insurance. Besides Hull & Machinery Insurance and Cargo Insurance offered on the commercial market in insurance, Protection and Indemnity Insurance is a ship owner's insurance cover for legal liabilities to "third parties" who are any person apart from the ship owner himself who may have a legal or a contractual claim against the ship "in operation."

Protection & Indemnity Insurance is usually covered by entering the ship in a mutual insurance club. Originally, the mutual aspect was obtained by simply grouping together owners who operated similar ships on similar trades. Now, mutuality is achieved by an underwriter who endeavours to see that owners carry their fair share of the risk.

The members of P&I clubs are ship owners, charterers or ship management companies. Legal liability is decided in accordance with the laws of the country where the damage occurred.

The protection and indemnity insurance cover for contractual liability is agreed at the time the owner requests the insurance cover from the club and is usually in accordance with the owner's liability under crew contracts or special terms related to the trading pattern of the vessel. The insurance is an indemnity type of insurance which means that the club member has the duty to demonstrate his loss (or damage) before the club pays out under the insurance policy.

The cover includes liability for accidents such as when the member's ship is in collision with another ship (one third liability) or when the entered ship strikes a fixed or floating object (quay, dock or buoy). However, collisions and striking liabilities are often covered under the ship's Hull and machinery cover. Protection and Indemnity insurance covers an owner's liability for all personal injuries which occur on board, including passengers, stevedores. Pilots and visitors to the ship. Protection and indemnity insurance also covers a ship owner's liability to pay for the cost of repatriating crew members who become or are injured on board, including an owner's expenses for hospital bills, the cost of repatriation of crew members, the costs of sending replacement crew to the ship if necessary and the owner's liability for loss of crew belongings in cases of shipwreck or fire on board. Other risks covered include liability for stowaways, liability related to drugs, smuggling, liability for oil pollution and other types of pollution, legal liability for wreck removal if the ship sinks and is blocking free navigation for other vessels.

The author thanks The University of Edinburgh through the corporation "Edinburgh Research & Innovation", The World Maritime University (Postgraduate Research) and the Joint LLM Programme WMU-Lund University (Sweden) in International Maritime Law for kind suggestions which led to the publication of the book.

2. Focus of the study

At present, a major function of Protection and Indemnity Insurance is to cover the ship owner for legal and contractual liability for loss of cargo or cargo damage if there has been a breach of the contract of carriage. This liability is called "Third party liability" and is most frequent for a ship owner.

In practice, damages to cargo are often caused by small mistakes. The owner of the cargo will recover his losses for loss or damage to cargo from the cargo underwriter who may lodge a claim directly with the ship owner. He may also sell the claim to a professional claim recovery agent. The ship owner will usually hand over the handling of the cargo claim to his P&I club. The claimant will often allege that the ship was unseaworthy at the commencement of the voyage. The ship owner has to prove that the ship was seaworthy. Failing to do so, the ship owner loses the opportunity to invoke the 17 exculpatory clauses which are usually incorporated with the bill of lading. The exculpatory clauses according to the Hague Visby rules are listed under Article 4, Paragraph 2.

The claim handling process is often lengthy and complicated. One of the difficulties is about collecting the evidence in order to establish the exact cause of the damage or loss of cargo. This can be done by referring to the bill of lading: An important function of this document is to describe with accuracy the condition and quantity of the cargo as received on board. If the cargo is discharged in a different condition or in lesser quantity than that entered on the bill of lading, the ship owner will be held liable for the damage or shortage. In order to clarify this ship owner liability, Article 3, Paragraph 2 of the Hague Visby rules stipulates:

"Subject to the provisions of Article 4 ..., the carrier shall properly and carefully load, handle, stow, carry, keep, care for and discharge the goods carried."

The focus in this study is placed on the function of Protection and Indemnity insurance covering the ship owner's liability for damages to cargo.

The present liability regimes will be scrutinized. As far as the legal regimes are concerned, following issues will be addressed.

What is the legal nature of the bill of lading and what is its role in Protection and Indemnity Insurance?

Damage to the Cargo: What is it and how can it be measured?

What is the nature of the contract of carriage? What about the rupture of the contract?

How can the applicable law be defined in case of loss or damage to cargo?

What party bear the burden of proof and what is the order in the administration of proof for liability for cargo damage?

In addition to studying liability regimes, the work would show how protection and indemnity insurance creates the needed balance and stability in the international seaborne trade through the concept of mutuality.

Eventually, an outlook on the future of Protection and Indemnity Insurance with emphasis on prevention for cargo damage or loss will be drawn. The main practical interests in conducting this legal study can be summarized in the following assertions.

*Emphasis in the study will be placed on practical cases: Issues of practical relevance requiring a careful analysis in liability for cargo claims will be dealt with.

* Since cargo liability is now a major area of ship owner's liability for third party considering the number of claims and the operating cost for the ship owner in spite of progress in technology and improvements in safety standards in shipping, there is a need for a comprehensive study leading to appropriate measures which should protect the financial stability of the ship owner and prevent damages to cargo or losses of cargo.

*In addition to that, this study would be a useful contribution in the area of marine insurance where the daily work in claim processing depends very much on the reputation and the size of the P&I club and the personal experience of the claim officer. Studying practical cases and reviewing well established practice at the 13 P&I clubs members of The International Group in the area of cargo claims would help people involved in international maritime trade improve their skills and serve as bench mark in the branch.

*Finally, and in combination with the academic need, this work would achieve a well needed balance in the academic world between the theoretical and systematic research from universities and the available and well needed but less known empirical knowledge accumulated in international companies involved in international maritime trade.

The undertaken work is essentially based on court decisions, arbitration awards, practice at selected P&I clubs, comprehensive and systematic review of relevant research (available literature and internet sources) and views expressed by experts in the field of marine insurance and international maritime trade.

Eventually, carefully selected statistics will illustrate the facts with reference to the importance and the value of concepts in maritime transport; to some extent, the work would be enriched by the author's professional experience as Claim Executive Officer in charge of P&I insurance matters and Head of Strategic Logistics dealing with export of electronics. Including dehumidifiers and radar cassettes.

3. Relevance of the project to the Postgraduate Programme in Maritime and International Commercial Law

I believe that a strong link between the project and Postgraduate Studies could be found at least in the following research areas: research in maritime law and marine insurance law.

This book is about practical legal issues related to the ship owner's liability for cargo claims and the protection offered by P&I clubs in this respect.

Ship Owner's Liability for Cargo Claims

Ship owner's liability for cargo claims has steadily been increasing. This liability means that a damage to the cargo or loss of cargo occurred while the cargo was on board the ship which the owner could be held liable for. The cargo underwriter will normally pay the cargo owner. The cargo underwriter will then seek to recover his losses from the ship owner through subrogation. One of the following international conventions determines the applicable regime of the ship owner's liability:

- Hague Rules (1924)
- Hague Visby Rules (1968) and
- Hamburg Rules (1978)

Cargo Claims in Protection and Indemnity Marine Insurance

As underlined previously, at present a major function of Protection and Indemnity Insurance is to cover the ship owner for liability for loss of cargo or damage to cargo if there has been a breach of the carriage contract.

P&I Insurance is as important to a ship owner as his Hull & Machinery insurance is or Cargo insurance to a cargo owner engaged in international maritime trade. Special attention would be given to the following issues related to P&I insurance:

- The concept of mutuality as a key to the club system success,
- Showing how the concept has evolved taking into account new trends in international shipping,
- The role of reinsurance as financial stability factor in Protection and Indemnity Marine Insurance.

INTRODUCTION

The carriage of goods by sea is mainly regulated by the following international conventions:

*Hague Rules (1922),

*Hague-Visby Rules (1968) and

*Hamburg Rules.

These conventions have generally been incorporated in national legislations which might be slightly different from the international conventions in some respects.[1]

While addressing the issues pertaining to the legal regimes associated with the carriage of goods by sea, it is worth mentioning that there are quite many publications on the subject. Nevertheless, only a

---

[1] Gold (Edgar), *Gard Handbook on P&I Insurance,* 5th Edition, 2002, p.p. 332-357; Tetley (William), "Interpretation and Construction of the Hague, Hague-Visby and Hamburg Rules" (PDF9), 2004, 10 JML, p. 30-70; Max Planck Institute of Foreign Private and Private International Law, "Maritime Transportation (being Chapter 4 of Volume XII), Hamburg 2004

few of them could be connected to the real practice of maritime law and practice. Some of them presenting the subject quite similarly. Needless to state that they have the same starting point, namely: the content of the international maritime conventions.

In this dissertation, an analytic and selective approach has been applied in order to assess the main legal issues according to the following criteria:

*Practical relevance,

*The main lessons learned from outstanding academics with interests in the International Maritime Law and Marine Insurance,

*My professional observations showing how the international conventions mentioned above are actually being interpreted and applied in Maritime Law and Practice,

*Specification terms in the Hague Rules, Hague-Visby Rules and Hamburg Rules.

Special attention has been given to the following scholarly publications for a deeper insight in the maritime conventions named above:

De Wit (Ralph), *Multimodal Transport: Carrier Liability and Documentation,* Lloyd's of London 1995 (including Supplement, 2002)

Gold (Edgar), C.M., Q.C., Chircop (Aldo), Kindred (Hugh M.), *Maritime Law,* 2004

Gold (Edgar), *Gard Handbook on P&I Insurance,* 5th Edition

Hoek (Marianne), *Multimodal Transport Law: The Law applicable to the multimodal contract for the carriage of goods, 2010*

Kouladis (Nicholas), *Principles of Law Relating to International Trade,* New York Inc, 2006

Reynold (Francis), "The Hague Rules, The Hague-Visby Rules and The Hamburg Rules,"; In MLAANZ JOURNAL, (1990), 7, p.19

Sampani (Constantina), Aberystwyth University & Mr Stephen Wrench, Lloyd's Maritime Academy, Informa plc, 'Reflexions on Legal Issues in a Diverse Maritime World', Cambrian Law Review, Volume 44, 2013

Sturley (Michael F.), ed., "The Legislative History of the Hague Rules," Vol.2, F.B. Rothn Littleton, Colorado, 1990

Tetley (William), *Marine Cargo Claims,* Vol. I and II, Thomson*Carswell, 2008

Tetley (William), "The Proposed New United States Senate COGSA, COGSA 98, The Disintegration of Uniform International Carriage of Goods by Sea Law," (199) 30, Journal of Maritime Law & Commerce, 595-625

Watson (Alasdair), *Finance of International Trade,* 7th Edition

Wilson (John F.), *Carriage of Goods by Sea,* 3rd Edition

CHAPTER I

HAGUE RULES

1.1. ORIGIN

These rules are officially known as "International Convention for Unification of Certain Rules of Law in Relation to Bills of Lading and Protocol of Signatures." They established standard basic obligations and responsibilities of the shipper and the maritime carrier for goods covered under a bill of lading, The Hague Rules are sometimes referred to as "The 1924 Brussels Convention."

The main problem to be solved at that moment was the widespread dissatisfaction among cargo owners and their insurers with arbitrary restrictions imposed by maritime carriers in order to limit their liability in case of loss or damage to cargo.

After some changes, the United States of America adopted these rules in 1936 as Carriage of Goods by Sea Act or COGSA, in short form.[2]

The Hague Rules were elaborated at a meeting of the Maritime International Law Association, (CMI, Comité Maritime International) at the Hague in the Netherlands in 1921. They were finally adopted at a diplomatic convention in Belgium (Brussels 1924). That is why they are also referred to as the Brussels Convention, although they are normally called "The Hague Rules."

The Hague Rules are very much connected with The Netherlands and Belgium. They represent the first effective internationally agreed control of bill of lading terms leading to a large unification of rules relating to the carriage of goods by sea under the bill of lading. [3]

1.2. HAGUE RULES LIABILITY SYSTEM

The liability system is mainly concerned with physical damage, damage due to machine breakdown, deficient ventilation as well as delay in goods delivery. There lies upon the carrier a duty to make the vessel seaworthy. The concept of seaworthiness can be described as follows:

---

[2]   Tetley (William), "Interpretation and Construction of the Hague, Hague-Visby and Hamburg Rules", (PDF), MCgill, Maritime Law, 2008, p. 43

[3]   Tetley (William), "Interpretation and Construction of the Hague, Hague-Visby and Hamburg Rules", MCgill, Maritime Law, 2008, p.p. 30-70

*Seaworthiness from the technical point of view (Ship's design: Hull and Machinery; Conditions and ship's stability in fitness for the carriage of goods);

*Cargo worthiness (the required ship for a given cargo, for instance a refrigerated container ship for the carriage of meat);

*Seaworthiness for the intended voyage (the ship should be fitted according to the geographical and weather conditions in the trading areas).

Therefore, the carrier shall be bound before and at the beginning of the voyage to exercise "due diligence" in order to:

a)   Make the ship seaworthy,

b)   Properly man, equip and supply the ship,

c)   Make the holds, refrigerating and cool chambers, and all other parts of the ship in which goods are carried, fit and safe for their reception, carriage and preservation."

It would be appropriate to mention that there is no absolute liability on the carrier with regard to vessel seaworthiness, but he has to "exercise due diligence."

One of the principal rules in the Hague Rules liability is the following: "Neither the carrier nor the ship shall be responsible for loss or damage arising or resulting from:

...

(q) Any other cause arising without the actual fault or privity of the carrier, or without the fault or neglect of the agents or servants of the carrier, but a burden of proof shall be on the person claiming the benefit of this exception to show that neither the actual fault or privity of the carrier nor the fault or neglect of the agents or servants of the carrier contributed to the loss or damage." [4]

The carrier is liable for loss or damage arising from his own fault or privity or his servants or agents fault or privity. The burden of proof is on the carrier; this means that he has to prove that he, his servants or agents have not caused the loss or damage by negligence. There are a number of exoneration causes which relieve the carrier from his liability for loss or damage resulting from:

(c) Perils, dangers and accidents of the sea or other navigable waters.

(d) Act of god,

(e) Act of war,

(f) Act of public enemies,

(g) Arrest or restraint of princes, rulers or people, or seizure under legal process,

(h) Quarantine restrictions,

(i) Act or omission of the shipper or owner of the goods, his agents or representative,

(j) Strikes or lockouts or stoppage or restraint of labour from whatever cause, whether

---

[4] Ihre (R), Gorton (Lars) & Sandervärn (A), *Shipbroking & Chartering Practice*, 2nd Edition, p.p. 57-72

    Partial or general,

(k) Riots and civil commotions,

(l) Saving or attempting to save life or property at sea,

(m) Wastage in bulk or weight or any other loss or damage arising from inherent defect,

    Quality or vice of the goods,

(n) Insufficiency of packing

(o) Insufficiency or inadequacy of marks and

(p) Latent defect not discoverable by due diligence."

Actually, these exemptions illustrate situations where the carrier is not negligent. [5]

They are similar to the Civil Law concept known as "cas de force majeure." However, there are two "true" exceptions to the principal rule, namely: "error in navigation" and "fire."

With regard to the error in navigation, it is worth stating that only negligence in the navigation or the management of the ship can relieve the ship owner from liability.

Eventually, The Hague Rules do not apply to contracts of carriage under which live animals have been carried and cargo has been carried on deck, and the bill of lading states so expressly.

---

[5]   Ihre (R), Gorton (Lars) & Sandervärn (A.), Op. Cit., p. 61; Reynolds (Francis), "The Hague Rules, The Hague. Visby Rules and The Hamburg Rules", MLAANZ Journal, (1990), 7, p. 19

1.3. MAIN FEATURES OF THE HAGUE RULES

The main provisions in the rules are presented against the operational background at the time of their introduction. We have got to learn that many ship owners were at that time undertaking no liability at all. A famous quotation from the Annual Report of West of England P&I Club in 1889 was stating as follows: "The Committee congratulates the members on the absence in recent years of cargo claims which has been brought about by the now general adoption of the negligence clause, the premium reduction for use of this clause is therefore discontinued." This is the illustration of the reality at that time. [6]

In certain areas, especially the North Atlantic, some ship owners were excluding virtually all or a great deal of their liability.

1.3.1. SCHEME OF SPLIT RISK

The maritime carrier has rather a duty of reasonable care with regard to the ship seaworthiness and the care of cargo. Shippers' risks are negligence in navigation and management. The idea behind that was that sea transit was a dangerous adventure: Anyone participating in it assumes that the carrier will do the best he can and therefore, it is fair to excuse him of the particularly maritime, as oppose to the bailee aspects of the responsibilities undertaken. [7]

---

[6] Reynolds (Francis), "The Hague Rules, The Hague-Visby Rules and The Hamburg Rules", MLAANZ Journal, (1990), 7, p, 17

[7] Reynolds (Francis), Op. Cit., p.18

The Harter Act compromise (England) was taken through to The Hague Rules representing a compromise between ship owners, mainly British, who operated under considerable immunities which they were not willing to surrender, and cargo-owning countries, such as USA and Canada.

The rules are apparently a scheme for uniformity of bills of lading adopting the Harter Act compromise of the split risk between carriers' risks and cargo owners' risks.

Finally, on the whole, the balance seems to be in favour of cargo interests, bearing in mind that the ship owner was not allowed to exclude his liability beyond the rules provisions. This means that the ship owner could not exclude his liability for due diligence as to seaworthiness and care of cargo even if he has in return exoneration for liability in case of negligence in navigation and error in management.

This could be considered as a compromise leaning in favour of cargo owners. However. One has to remember that the ship owner obtained through the convention the benefit of the one year time bar and the package or unit limitation representing considerable benefits to the ship owner.

One hundred pounds sterling gold value in 1924 is, nevertheless, the equivalent of a fairly considerable amount of money.

That is what shows cases like "The Rosa S2" and "Brow Boveri versus Baltic Shipping Company" (Australia).

Further reading is provided in the following articles: 1988 2 Lloyd's Rep. 574 and 1989 1 Lloyd's Rep. 518.

CHAPTER II

HAGUE-VISBY RULES

1. BACKGROUND

Applying the Hague Rules to outward shipments for countries who were parties to the convention raised some practical problems regarding the law applicable to the contract: "The proper law of the contract." [8]

In addition to that, the question raised by the "Paramount Clause" in the bill of lading has to be dealt with.

What happens if litigation arises on shipments from country X in country Y?

The expectation was to make sure that the courts of country Y applied the Hague Rules

---

[8] Reynolds (Francis), Op. Cit., p.20

by using the paramount clause which requires the application of the Hague Rules to such shipments. A difficulty might arise when the clause paramount is omitted.

Let us illustrate the situation with a practical case.

Case: "The Vita Food litigation"

The case is related to a shipment out of Newfoundland, which was a convention country at that time independent. The case had to be settled in Nova Scotia. The ship belonged to a Nova Scotia corporation. Nova Scotia was also a convention country.

The clause "Paramount" was omitted from the bill of lading. It was not even really clear that the rules were compulsorily applicable to a shipment out of a convention country if the bill of lading contained a choice of law for another jurisdiction.

The now known as "Vita Food" gap enabled the rules to be evaded by a choice of law clause for a jurisdiction which either had not adopted the rules at all or did not apply them to the voyage.

The second subject to be dealt with was the result of an English case called Scrutton versus Midland Silicones.

The case held that the protection of the Hague Rules did not affect stevedores, since they were not parties to the contract of carriage, so that although you could not sue the carrier, you might well be able to sue the stevedoring firm and that firm could not rely on the time bar or the package or unit limitation the ship owner has right to. In order to get round this difficulty, attention was given to drafting the "Himalaya" clause, called so because it was drafted to deal with the problem first manifested in an earlier court of appeal case called "Adler versus Dickson."

Mrs Adler was a passenger on the P & O liner "Himalaya."

She could not sue the ship owner.

Thirdly, the case "Muncaster Castle" held that the ship owner's duty as to due diligence in furnishing a seaworthy vessel was non-delegable. This means that the ship owner could not argue that he had exercised due diligence by appointing competent maritime surveyors or repairing companies and similar actions. If those organisations were themselves negligent, the ship owner should be held liable. Nevertheless, carriers thought that they should be able to discharge their duty of care by delegating it to knowledgeable and skilled independent contractors.

Fourthly, the issue pertaining to the probative effect of bills of lading had to be examined. There were questions of exactly what statements in the bill of lading were proof, and to what extent they could be disproved – statements as to the amount of goods loaded in apparent order and condition on shipment.

Fifthly, the problems regarding the package or unit limitation due to inflation had to be discussed. The limit was 100 pound sterling, "taken to be" gold value, but no one knew the real meaning.

Actually, they were different values of gold, two tier systems and so on.

The remaining points were worked on and a draft was produced at the CMI Conference in Stockholm in 1963 and signed at the City of Visby on the island of Gotland in the Baltic at the end of the conference. Further work was done on the document and, extensively amended, it became the subject of an International Protocol to the 1924 Convention, and the rules as amended are called the Hague-Visby Rules which came into force in 1977.

The Hague-Visby Rules are simply the Hague Rules with a fairly small number of alterations, some of them quite important. [9]

They are Hague Rules with certain alterations made in the interest of correcting particular difficulties perceived at that time as having emerged from the application of the Hague Rules.

2. HAGUE-VISBY RULES VERSUS HAGUE RULES

Is the content of the Hague-Visby Rules different from the Hague Rules?

In the former the procedure for "internationalisation" is more efficient (in common law countries) if the domestic legislation bringing the rules into effects is properly drafted.

---

[9]   Reynolds (Francis), `The Hague Rules, The Hague-Visby Rules and The Hamburg Rules`, MLAANZ Journal, 1990, p.22

Legal Case: *"The Vita Food Gap"*

Article X of the Hague-Visby Rules contains provisions dealing with the application of the rules.

In England there is a strong case on the subject, "The Hollandia." [10]

The facts reveals that there was a shipment out of Scotland on a Dutch vessel. The bill of lading was governed by the Dutch law.

In the Netherlands, they had the Hague Rules providing for lower package or lower unit limit. According to the bill of lading, the Dutch law should apply and any litigation should take place in Holland.

The English court found the jurisdiction clause in the Netherlands as reducing the carrier's liability under Article III. 8 and declared it voided.

The Dutch jurisdiction clause was therefore struck out and the issue became simply one of "forum conveniens" leading to decision in favour of English jurisdiction. Besides, the jurisdiction issue, the question was raised whether stevedores who are normally considered as independent contractors should be entitled to avail themselves of the defences and limits of liability according to the Hague Rules whereby the servant or agent not being an independent

---

[10]   Hollandia Case, The (1983) IAC: 565 349; Gold (Edgar), *Gard Handbook on P&I Insurance*, p. 43

contractor shall be entitled to the ship owner's defences and limits of liability.

On one view, it could be meant to cover stevedores and some special interpretation functions should be deployed to try and make sure they come under the words "servant or agent."

On another view, international agreements in Brussels in 1968 was not secured to the exemption of independent stevedores (statement of Mr Justice Murphy in the High Court of Australia in the New York Star, 1978 139 CLR 231, 285).

The second problem was the meaning of the provisions stated in Article IV Bis.

ARTICLE IV BIS 1

"The defences and limits of liability provided for in these Rules shall apply in any action against the carrier in respect of loss of or damage to the goods covered by a contract of carriage whether the action be founded in contract or in tort."

It is not clear whether actions in tort against the ship owner are covered in the Rules. The matter is almost untested, something leading to a situation of controversy.

The view that the provision covers actions in tort against the carrier "in certain circumstances" was defended by Mr Anthony Diamond, QC, a well-known barrister and writer in the area of carriage of goods by sea (In 1978 LMCLQ, pages 248-253).

However, later another strong opinion on the issue was given by the English Court of Appeal in a case known as "The Captain Gregos."

The court took the narrow view that the provision only protects carriers who are sued in tort when they could have been sued in contract (1990 1 Lloyd's Rep. 310). An appeal to the House of Lords was unlikely.

Nevertheless, the matter cannot be considered as finally settled.

The third problem, namely that associated with "The Muncaster Castle" was abandoned at some stage in the negotiations.

The fourth problem was related to the bill of lading conceived as the receipt of the goods carried on the ship. This is dealt with by the addition of one sentence in Article III which requires the carrier on demand to issue a bill of lading containing certain particulars.

Article III. 4 then clarifies as follows: "Such a bill of lading shall be prima facie evidence of receipt by the carrier of the goods therein described in accordance with paragraphs (a), (b) and (c)."

That, together with one more sentence written in the Visby Rules becomes more significant. [11]

---

[11] Gold (Edgar), *Gard Handbook on P&I Insurance,* 5th Edition, Arendal, Norway, 2002, p.p. 332-357

It is provided in the Hague-Visby Rules the following: "Proof to the contrary shall not be admissible when the bill of lading has been transferred to a third party acting in good faith."

The provision is well known and generally applied in civil law countries. Nevertheless, it has also effect in common law countries because it potentially provides a solution for similar cases to that of "Grant v. Norway." [12]

Hence, the new provision stating that a master has no authority to sign a bill of lading for goods not on board, so that if the goods are not actually on board when the bill of lading is signed, the bill of lading does not bind the ship owner. But, there might be some slight technical difficulties in it. Here is the wording of the rule: "Proof to the contrary shall not be admissible ..."

Firstly, the wording is not clear. The rule makes the bill of lading prima facie evidence of receipt of goods by the carrier. This is to say that the owner cannot prove that the goods were not received when the bill of lading had been issue.

The second difficulty is that if no goods were ever shipped under no contract of carriage at all then the Hague-Visby Rules would not be applicable.

In addition to that, the question of package or unit limitation was dealt with. The limitation was intended to provide for the inflation

---

[12]   (1851) 10 CB 665, 138 ER 263 20 LJCP 93, 15 JUR 296

issue by linking the package or unit limitation to the "Poincarré franc", standing for a unit of currency defined by gold content.

Afterwards, this solution was not very satisfactory and the limit is now defined by SDR standing for Special Drawing Rights on the International Monetary Fund (IMF). In practice the limit was raised (to more than 100 pounds sterling).

Moreover, there was concern about bulk cargo. That is why a provision as to weight was added. Some countries wanted weight to be the only criterion for the package or unit limitation.

An alternative solution was added to the Hague Rules under Rule IV.5 (c) of the Hague- Visby Rules

"...*Neither the carrier nor the ship shall in any event become liable for any loss or damage to or in connection with goods in an amount exceeding 666.67 units of account per package or 2 units of account per kilogram of gross weight of the goods lost or damage, whichever is the higher*

Finally, there is a rule dealing with the container provision. The question was raised in order to determine whether the container is the package or unit, or the contents (of the container). The answer depends on the wording in the Bill of Lading.

Article IV. 5 (d), Hague-Visby Rules,

"... Where a container, pallet or similar article of transport is used to consolidate goods, the number of packages or units enumerated in the bill of lading as packed in such article of transport shall be deemed the number of packages or units for the purpose of this paragraph as far as these packages or units are concerned."

The wording "One container said to contain machinery" in a bill of lading means the container is the package or unit whereas the wording "One container containing 600 TV sets" means the sets would be the packages or units.

Concluding the chapter dealing with the Hague-Visby Rules attention has to be drawn to Article IV. 5 (e).

"...Neither the carrier nor the ship shall be entitled to the benefit of the limitation of liability provided for in this paragraph if it is proved that the damage resulted from an act or omission of the carrier done with intent to cause damage or recklessly and with knowledge that damage would probably result."

Hence, the package or unit limitation is broken if the ship owner acts with intent to cause damage or recklessly.

The importance of the provision in civil law countries is obvious. It is intended to make the ship owner act more carefully and to avoid maritime fraud.

Failing to do so, he would lose the compensation he might be entitled to through the P&I insurance policy.

Cargo interests have been considering both the Hague Rules and the Hague-Visby Rules as being largely in the interests of ship owners and ship owning countries.

It was in this context that the UNCTAD secretariat started to consider works in order to strengthen cargo interests and maritime related business in developing countries.

The final result was the Hamburg Rules. [13]

---

[13] Reynods (Francis), 'The Hague Rules, The Hague-Visby Rules and The Hamburgh Rules', MLAANZ Journal, (1990), p. p 29-34

CHAPTER III

Hamburg Rules

The rules originated from a report written by the UNCTAD secretariat around 1970. The document drew attention to certain issues associated with the application of the Hague Rules and the Hague-Visby Rules which were considered to be disadvantageous to cargo-owning countries and to developing countries particularly because of the fact that they led to the establishment of more business in developed countries (most of them being maritime countries, protecting ship owners, and also creating double insurance situations where the cargo owner was actually bearing the cost of insurance for liability of the ship owners.[14]

---

[14] Reynolds (Francis), 'The Hague Rules, The Hague-Visby Rules and The Hamburg Rules', MLAANZ Journal, (1990), p. 29

1. SOME DIFFICULTIES PERCEIVED IN THE HAGUE RULES AND THE HAGUE-VISBY RULES

1.1. Regarding the expected perils in Hague and Hague-Visby Rules, some of them were conceived as shipping was in its early days. [15]

1.2. When does the application of the rules start and when does it stop?

It is not clear whether the rules come into operation on the ship's rail when crossing the ship's rail or not.

Actually, the rules themselves say in article II:

"… Under every contract of carriage of goods by sea the carrier in relation to loading, handling, stowage, carriage, custody, care and discharge of such goods shall be subject to the responsibilities and liabilities and entitled to the rights and immunities hereinafter set forth."

That means that the carrier has to load and unload as part of his functions, even though charging and discharging are usually performed by stevedoring companies.

The English view on the matter is more realistic. It was established in the case "Pyrene versus Scindia" (1954 2 QB, 402) that it depends on the particular contract as to how much loading and unloading the carrier does.

---

[15] Reynolds (Francis), Op. Cit., p.p. 30-34

The contract can determine to what extent the rule shall apply.

Some authors refers to the "Concept of tackle-to-tackle period" meaning the liability period of the maritime carrier is the time under which the goods are under the actual control of the ship owner.

My personal opinion is that reference to the INCOTERMS should actually be the appropriate solution. Some scholars might argue against this opinion by reminding that the INCOTERMS are associated with the relationship between sellers and buyers in International Trade. This seems an artificial view on the issue due to the fact that more often the sellers as manufacturers arrange the overall transportation chain with due regard to the buyers. These parties are in principle in a long term commercial relationship (Edinburgh, Law School, PGCLDG 2012).

1.3. There are certain matters hardly touched at all, for instance, the delay in goods delivery.
1.4. Problems associated with the time bar: Why should there be a short time bar for actions against the ship owner?
1.5. What is the meaning of the package or unit limitation? Why it exists at all?
   And, if the provision is needed, why is it so low?
1.6. There are shortcomings in the Rules about jurisdiction and arbitration clauses and this allows carriers to require arbitration and/or litigation in countries of their choice.

The general opinion was that this led to disadvantages to developing countries and to cargo-owning countries, because it was said to be more in advantage of carriers.

With regard to cargo-owning countries, it was argued that one might have to bear the cost of insurance which should have been paid by the ship owner bearing in mind the difficulty in establishing the liability of the carrier. It should be clarified here that the ship owner bear the cost of the insurance for his liability which is covered by his P&I Club whereas the consignee or cargo-owner insures the cargo through the cargo insurance policy available on the commercial market of insurance; it is important to make the difference between the two kinds of insurance.

Eventually, there were risks of loss or damage over and above the package or unit limitation, of missing the time bar and so on.

2. MAIN DIFFERENCES BETWEEN HAMBURG AND HAGUE-VISBY RULES

The principal feature of the Hamburg Rules is the new basic rule of liability in Article V.

2.1. "*The carrier is liable for loss resulting from loss of or damage to the goods as well as from delay in delivery, if the occurrence which caused the loss, damage or delay took place while the goods were in his charge as defined in article 4, unless the carrier proves that he, his servants or agents took all measures that could reasonably be required to avoid the occurrences and its consequences.*"

A new joint and several liability of carriers is then established in the Hamburg Rules. [16]

This liability rule makes it possible to sue the actual carrier when a charterer's bill of lading is issued. It is then easier to sue the actual carrier, and also for the actual carrier to rely on the excepted perils

Moreover, there are rules about liability for delay with special limits on damages for it. The convention contains provisions dealing with deck cargo. It is specified when cargo can be carried on deck and the conditions associated with the carriage of live animals are stipulated.

The provisions apply to specified voyages. Actually, they even give indications on where litigation or arbitration may take place.

Finally, let us just look at the following provisions, Article 2.1.1. in the Hamburg Rules.

Article 2. 1.1. stipulates: "In judicial proceedings relating to the carriage of goods under this convention, the plaintiff, at his option, may institute an action in a court which, according to the law of the State where the court is situated, is competent, and within the jurisdiction of which is situated one of the following places:

    (a)   The principal place of business or, in the absence thereof, the official residence of the defendant; or

---

[16]   Reynolds (Francis), 'The Hague Rules, The Hague-Visby Rules and The Hamburg Rules', MLAANZ Journal, (1990), p. 30

(b)   The place where the contract was made …; or

(c)   The port of loading or the port of discharge; or

(d)   Any additional place designated for that purpose in the contract of carriage by sea."

The provision should reduce the effect of some carriers' jurisdiction clauses.

With regard to arbitration, following provisions specified in Article 22, Hamburg Rules, are worth mentioning.

Article 22, Hamburg Rules:

"Arbitration

1.   Subject to the provision of this article, parties may provide by agreement evidenced in writing that any dispute that may arise relating to carriage of goods under this Convention shall be referred to arbitration.

2.   Where a charter-party contains a provision that disputes arising thereunder shall be referred to arbitration and a bill of lading issued pursuant to the charter-party does not contain a special annotation providing that such provision shall be binding upon the holder of the bill of lading, the carrier may not invoke such provision as against a holder having acquired the bill of lading in good faith.

3.   The arbitration proceedings shall, at the option of the claimant, be instituted at one of the following places:

a) A place in a State within whose territory is situated, meaning:

(I) The principal place of business of the defendant or, in the absence thereof, the habitual residence of the defendant; or

(II) The place where the contract was made, provided that the defendant has there a place of business, branch or agency through which the contract was made; or

(III) The port of loading or the port of discharge; or

b) Any place designated for that purpose in the arbitration clause or agreement.

4. The arbitrator or arbitration tribunal shall apply the rules of this Convention.

5. The provisions of paragraphs 3 and 4 of this article are deemed to be part of every arbitration clause or agreement, and any term of such clause or agreement which is inconsistent therewith is null and void.

Nothing in this article affects the validity of an agreement relating to arbitration made by the parties after the claim under the contract of carriage by sea has arisen."

LETTERS OF GUARANTEE

These are letters issued by shippers whereby assurance related to the accuracy of the cargo is given to the carrier. In case of no accuracy, the shipper will compensate the carrier.

The provisions on letters of guarantee which are dealt with at Article 17 (Hamburg Rules) are not popular in maritime countries because they could lead to maritime fraud.

Article 17

GUARANTEES BY THE SHIPPER

1. The shipper is deemed to have guaranteed to the carrier the accuracy of particulars relating to the general nature of the goods, their marks, number, weight and quantity as furnished by him for insertion in the bill of lading. The shipper must indemnify the carrier against the loss resulting from inaccuracies in such particulars.
   The shipper remains liable even if the bill of lading has been transferred by him. The right of the carrier to such indemnity in no way limits his liability under the contract of carriage by sea to any person other than the shipper.
2. Any letter of guarantee or agreement by which the shipper undertakes to indemnify the carrier against loss resulting from the issuance of the bill of lading by the carrier, or by a person acting on his behalf, without entering a reservation relating to particulars furnished by the shipper for insertion

in the bill of lading, or to the apparent condition of the goods, is void and of no effect as against any third party, including a consignee, to whom the bill of lading has been transferred.

3. Such letter of guarantee or agreement is valid as against the shipper unless the carrier or the person acting on his behalf, by omitting the reservation referred to in paragraph 2 of this article, intends to defraud a third party, including a consignee, who acts in reliance on the description of the goods in the bill of lading. In the latter case, if the reservation omitted relates to particulars furnished by the shipper for insertion in the bill of lading, the carrier has no right of indemnity from the shipper pursuant to paragraph 1 of this article.

4. In the case of intended fraud referred to in paragraph 3 of this article the carrier is liable, without the benefit of the limitation of liability provided for in this Convention, for the loss incurred by a third party, including a consignee, because he has acted in reliance on the description of the goods in the bill of lading."

Common Understanding in Hamburg Rule

At the very end of the Hamburg Rules a provision called "The common understanding" appears. It was introduced at the initiative of the United States. According to the clause, parties to the Convention agree with the idea that the carrier's liability is based on presumed fault, introducing a liability regime virtually similar to strict liability. [17]

RULES OF INTERPRETATION

As a general rule, Common Law countries have relied on and referred to judgments from the courts of England and the United States, while the doctrine, the laws and judgments of France have significantly influenced Civil Law jurisdictions in respect to the carriage of goods by sea. [18]

Continental authors, too, refer at times to English and American courts authorities in spite of limitations due to the language barrier.

It is argued that Maritime Law and the Law of Carriage of Goods by sea should be internationally uniform, because goods and ships move from one jurisdiction to another.

Merchants, shippers, consignees, carriers and underwriters could only have complete confidence in a contract if they are sure as to which law would apply and how it would be interpreted, no matter in what jurisdiction their claim or defence is heard.

This is called predictability or reliability of the rules of Law.

---

[17] Tetley (William), `Interpretation and Construction of The Hague, Hague-Visby and Hamburg Rules`, (PDF), 2004, 10 JIML, p. 71

[18] Tetley (William), `Interpretation and Construction of The Hague, Hague-Visby and Hamburg Rules`, 2004, 10, JIML, p.p. 50-70

The willingness of courts to consider the Hague and the Hague-Visby Rules with persuasive authority, and even to consider decisions of foreign courts on matters of interpretation of these conventions is greater today than in the past.

Occasionally, both The Hague or Hague-Visby Rules and a domestic statute can be considered together when solving a special issue. This is particularly true with regard to national bills of lading act or a national arbitration act dealing with a contract of carriage which is subject to the rules.

If a consistent and reasonable solution cannot be reached while applying the two statutes together, the latter statute is usually given supremacy. The Hague and Hague-Visby Rules are generally fair and commercially acceptable uniform international rules in the area of carriage of goods by sea leading countries to codifying statutes and providing an important link between the Civil Law and the Common Law, as well as being a fertile source of Comparative Law.

In particular, the jurisprudence relating to the rules illustrates that "interpretation" in Civil Law jurisdictions and "construction" in Common Law jurisdictions usually produce very similar results in practice. Such uniformity of interpretation is aided to a large extent by the understanding of these rules as an international instrument which requires a truly international interpretation so that they attain their purpose of harmonizing the International Law of Carriage of Goods by Sea.

Rules of Interpretation in Hamburg Rules

Whereas the Hague and Hague-Visby Rules reflect the Common Law style of precision with detailed drafting, the Hamburg Rules are drafted in the more concise, civilian style.

The Hamburg Rules contains a provision at Article 3 stating the following: "… It is obvious, that the Hamburg Rules being an international convention, should be interpreted to promote uniformity…"

This goes without saying; the rules have remained for most of the maritime countries as a single, non-mandatory rule of interpretation.

It is worth mentioning that provisions on "the common understanding" introduced in the rules at the request of the United States delegation introduces "the principle of presumed fault or neglect" for the carrier. The common understanding is not part of the Hamburg Rules; it seems to be an example of confusing drafting for people involved in the Carriage of Goods by Sea. [19]

The search for a better international convention led to the elaboration of a new legal instrument, the "Rotterdam Rules."

---

[19] Tetley (William), `Interpretation and Construction of the Hague, Hague-Visby and Hamburg Rules`, (PDF), 2004, 10, JIML, p. 71; Reynolds (Francis), Op. Cit., p. 31

CHAPTER IV

Brief Presentation of the "Rotterdam Rules"

INTRODUCTION

On the 21st of September 2010 a three- day signature ceremony began in Rotterdam for the Convention on Contract for the International Carriage of Goods Wholly or Partly by Sea. The convention is known as "The Rotterdam Rules."

It was adopted in December 2008 by the United Nations General Assembly and would enter into force twelve months after ratification by at least twenty States.

Signers included the United States of America, France, Greece, Poland, Denmark, Switzerland, Norway and the Netherland (Rotterdam, the 23rd of September 2009). Signatures were allowed after the signature ceremony at the UN Headquarters in New York City, New York.

The Rotterdam Rules are a set of international rules that revises the legal and political framework for maritime carriage of goods: they regulate the overall carriage of goods when the goods have been wholly or partly carried by sea. The Convention establishes a more modern, uniform legal regime governing the rights and obligations of shippers, carriers and consignees under a contract for door-to-door shipments that involve international sea transport.

The aim of the Convention is to extend and modernize international rules already in existence and achieve uniformity of admiralty law in the field of maritime carriage, hence updating and/or replacing many provisions in The Hague Rules, Hague-Visby Rules and Hamburg Rules. This is an ambitious undertaking with involvement of many experts in the field of Carriage of Goods by Sea, namely: maritime lawyers (CMI), Ship owners, shippers, marine insurers and maritime policy makers in order to achieve uniformity, standardization, modernization and fairness of the rules governing the International Carriage of Goods by Sea.

The process that has taken more than ten years of work might be finalized. However, a great deal of collaboration among all the actors with main interests in the international seaborne trade is still required.

In this short presentation of the rules, following issues will be dealt with:
 *The main features of the Rules,
 *The volume contract concept under the Rules,

*Arguments for the ratification of the Rules and
*Shortcomings/weakness of the ship owner's liability regime
under the "Rotterdam Rules."

The impact of the Rules on the Protection and Indemnity Marine
Insurance will receive a particular attention in the present work.

1. OVERVIEW OF THE ROTTERDAM RULES

Taking into account the changing patterns in International Trade,
the Rules recognize the application of the "volume contract
concept."

These rules are a positive innovation through the perceived efficient
balancing of risks between carrier and shipper interests: this is a
carriage regime focusing essentially on a balanced allocation of
obligation and liability between carriers and shippers.

Moreover, the rules govern the legal relationship between
carriers and shippers in terms of facilitating the free flow of trade
economically and efficiently.

It has to be noted that the freedom to derogate from the Rules with
regard to volume contracts is premised on the trade aspect of the
contractual relationship between carrier and shipper.

No derogation is allowed from those of the liabilities and obligations
of the Rules that are of the essence of all carriage conventions; and
this principle applies also to the Rotterdam Rules.

Notable provisions and changes of law found in the Convention are mentioned below.

*The Rules extend the period of time that carriers are responsible for goods to cover the time between the point where the goods are received to the point where the goods are delivered.

*They allow for more e-commerce and approve more forms of electronic documentation.

*They obligate carriers to provide seaworthy ships for the carriage of goods and properly crewed vessels throughout the voyage.

*They increase the limit liability of carriers to 875 units of account per shipping unit or three units of account per kilogram of gross weight.

*They eliminate the "nautical fault defence" which had prevented carriers and crewmen from being held liable for negligent ship management and navigation.

*They extend the time that legal claims can be filed to two years following the day the goods were delivered or should have been delivered.

*They allow parties to "volume" certain contracts to opt-out of some liability rules set in the convention.

The concept of "Volume Contract" needs some more precision in order to underline the singularity of the Rotterdam Rules with regard to carrier's liability.

2. THE VOLUME CONTRACT CONCEPT UNDER THE ROTTERDAM RULES

The opinion that the Rotterdam Rules have hurled us back to the chaotic situation of freedom of contract that prevailed before the Hague Rules with the provision relating to the volume contract appeared to be misconceived.

The freedom to derogate from the Rules based on volume contracts has to be considered as a mean to promote international trade in the contractual relationship between carrier and shipper.

The nature of this contractual relationship is of commercial nature.

3. ARGUMENTS IN FAVOUR OF THE ROTTERDAM RULES

The Rules offer the best prospects for updating the law associated with the carriage of goods by sea in a uniform approach across the world-

The International Chamber of Commerce threw its support behind the ratification of these rules for the following reasons:

*The convention would update the existing liability regimes as far as the Carriage of Goods by Sea is concerned,
* The proposed regime would underpin an important element in the International Seaborne Trade (Volume Contracts).

The American Bar Association well known for high legal standard has also urged the Congress to take the necessary steps for the ratification of the Rules arguing as follows:

"The present regimes are numerous and outdated…

The Rotterdam Rules will provide greater harmony, efficiency, uniformity and Predictability in the areas involved in maritime transport…"

Eventually, the Rules seem to have gained the support of the Maritime Industry, the International Community (through all the negotiation process), including the World Shipping Council which is a prominent supporter of the Rotterdam Rules.

4. THE PERCEIVED SHORTCOMINGS IN THE ROTTERDAM RULES

The Rules might contain some imperfections, but even critics of the adopted Convention recognize many positive innovations to the present liability regimes. Among the critical opinions the Belgian and Canadian views are mentioned here due to the particular importance of the transit carriage passing through these countries. It is known that a large amount of cargoes to be delivered in western European range would transit through Antwerp whereas the Saint-Laurent Seaway in Canada plays an important role in the transportation industry.

According to the Belgian Maritime Law Association, the freedom of contract provisions for volume contracts does not offer sufficient protection to smaller shippers and freight forwarders, who may be forced into contracts offering less legal certainty than the Hague-Visby Rules.

Even the provisions on the shipper's liability to the carrier are criticized for the fact that the shipper, in stark contrast to the carrier, is subject to unlimited liability, although shipping interests remarked that the shipper is intimately familiar with the nature of its cargo.

Regarding the jurisdiction rules, it was felt that the Belgian courts in particular would continue applying the EU Regulation 44/2001. The opting out clause in Article 74, combined with the separate status of regional integration described in Article 93, would probably render the jurisdiction provisions useless in Belgium.

Another critical opinion was expressed by the Canadian delegation. While many Canadian stakeholders indicated that Canada should sign the Convention, subject to ratification, there were also many stakeholders who could not support Canada's signature at that time and felt that such a step should be considered as and when Canada's major trading partners have indicated their commitment to ratify the new Convention.

Taking into account the diverse views among stakeholders on the Rotterdam Rules and the need to undertake further consultations mostly on the provisions related to domestic carriage of goods by water, Canada was therefore not in a position to sign the new Convention in Rotterdam on the 23rd of September 2009.

A more comprehensive opinion was expressed by Tetley, a legal scholar with good insights in the law associated with the International Carriage of Goods. He recognizes that the Rules are an attempt to find uniform standard rules in the Carriage of Goods which contribute to enhance fairness in the International Trade. Accordingly, it might be a good idea to leave intact areas of Carriage of Goods by Sea that work well, for example obligations and liability of the carrier as defined both in the Hague and in the Hague-Visby Rules. Maritime lawyers, the Maritime Industry and the International Community should look for new provisions to govern the areas needing attention.

It is well known that the general opinion is not in favour of the Hamburg Rules; countries with great maritime powers and the Maritime Industry find the Hague and the Hague-Visby Rules more suitable to the Carriage of Goods by Sea. The Rules need to be improved taking into account the advantage of globalization, the development in information technology and the trends in the International Maritime Industry.

William Tetley maintains that the UNCITRAL Draft Convention might be just recommendation rules in the future according to present researches in International Maritime Law: it fails to properly address multimodal carriage of goods given its maritime nature, and it goes too far in "disturbing" the existing established carriage of goods by sea regime.

Thus, leaving the UNCITRAL Draft Convention for later adoption, it would be advisable for the international community together with the United States where the Hague Rules are still in application to adopt the 1980 Convention related to Multimodal Carriage of Goods.

Meanwhile, specific issues on Carriage of Goods by Sea would need to be addressed by the further adoption of a protocol to the Hague-Visby Rules or a modified version of the Hamburg Rules. Such an instrument would essentially:

* add a definition of actual carrier that includes any maritime performing party (for instance, the concept of NVOCC);
* delete the outdated nautical fault defence;
* add rules governing deviation and deck carriage as necessary provisions;
* maintain the Hague-Visby Rules limits of liability and
* add provisions enabling the use of electronic documents.

In the context of the discussion above, the following factors could be considered.

Most of the organizations with great interests in seaborne trade support the Rotterdam Rules. Among them, the following are worth mentioning: the United Nations Commission on International Trade, the CMI, the World Shipping Council, the American Bar Association, various maritime academic institutions and the International Group providing Protection and Indemnity Insurance.

Besides, the Rules deal with the liability regime for the carriage of goods; they are designed to replace the existing three conventions on the ship owner's liability regimes.

In addition to that, provisions relating to the International Multimodal Transport of Goods are included in the Rules.

Maritime Lawyers at the University of Southampton believe that the Rotterdam Rules represent the most "comprehensive overhaul" of the Law of Carriage of Goods by Sea in more than fifty years.

Concluding with the study supervised by Professor Mukherjee, it certainly appears that the Rotterdam Rules provide an opportunity to create a legal framework within which those trading through volume contracts can operate within a flexible liability regime allowing certain derogations for the sake of economic efficiency in the global trading community. This Convention even with its imperfections is an all-embracing, comprehensive and well-balanced regime for the International Carriage of Goods.

Its strength lies in its singularity designed to replace three conventions for one with the addition of another convention that never was (UN Convention on International Multimodal Transport of Goods, 1980).

Eventually, it is the expectation of the international community that this legal instrument would significantly contribute to the improvement of International Trade in this growing economic globalization.

Closing this chapter, these recent cases in the United States are worth mentioning. They show how the Rules would apply.

Case (1)

Norfolk Southern Railway Co. v. Kirby

The court applies the rules laid down in the "Carmack Case" where the railway transport is considered as part of the maritime carriage.

Case (2)

Kawasaki Kisen Kaisha, Ltd.v.Regal- Beloit Corp (Carmack Amendment)

CHAPTER V

IMPACT OF THE MARITIME CARRIER'S LIABILITY REGIMES ON MARINE INSURANCE LAW & PRACTICE

International trade plays a vital role in the global economy which is in a crucial need of ships and the overall shipping industry. Most of the products needed in our global economy are brought to their consumers by sea.[20]

Nevertheless, despite the technical progress in shipping, the advent of the cellular containership and progress with regard to safety of ships, including progress in the area related to the protection of the marine environment, many casualties leading to significant loss of cargo occur in a much wider proportion than in the early days of shipping. With this enormous amount of cargo being moved all over the world, it is not surprising that casualties at sea occur.

---

[20]   Chuah (Jason), *Law of International Trade*, Third Edition, Thomson *Sweet & Maxwell, p.p.: 21-26

INSURING A VESSEL LIABILITY

It is estimated that the world merchant fleet exceeds 85,000 vessels with a combined gross tonnage of more than 1 billion GT (Gross tonnage).[21]

Marine Cargo insurers provide cover for known quantifiable goods in favour of cargo owners.

Besides Cargo insurance, there is also a need for ship owners to be protected against liability for loss or damage to cargo in carriage of goods.

This protection is generally offered through the legal institution of Protection and Indemnity Insurance.

Shipping is still needed and this necessity for ships will continue to grow in the future.

Therefore, the global insurance industry is necessary in order to help ship owners mitigate their risks and manage efficiently their ships.

---

[21] IGP&I (International Group of P&I Clubs), ANNUAL REVIEW 2011/2012, page 3

The basic provisions in carriage of goods by sea are formulated at article 3 of the Hague-Visby Rules which is the equivalent of article 3 in Hague Rules.[22]

Article 3, paragraph 2, of Hague-Visby Rules:

> *"Subject to the provisions of Article 4 ..., the carrier shall properly and carefully load,*
> *Handle, stow, carry, keep care for and discharge the goods carried."*.....

This formulation was adapted at different periods of time in order to meet the ever changing conditions in carriage of goods from bulk shipping to cellular container ships through the general cargo shipments context.[23]

---

[22] Tetley (William), *MARINE CARGO CLAIMS,* Fourth Edition, Volume 1, THOMSON *CARSWELL,2008, p. 6;

The Hague Rules stand for "International Convention for the Unification of Certain Rules of Law Relating to Bills of Lading, signed at Brussels, August 25, 1924 and in force as of June 2, 1931;

The term "Hague-Visby Rules 1968" refers to the Hague Rules 1924, as amended by the "Protocol to Amend the International Convention for the Unification of Certain Rules of law Relating to Bills of lading, adopted at

Bruxelles, February 23, 1968, which Protocol entered into force June 23, 1977

[23] Tetley (William), *Op. Cit.*, p.p. 6-91 (Hague, Hague-Visby Rules and Hamburg Rules)

Failure for the maritime carrier to fulfil his main duties of taking care of cargo once the goods are loaded on the ship and delivering the goods in the same state and quantity as the goods loaded leads to the carrier being held liable in case of damage or loss to cargo according to the Hague Rules, Hague-Visby Rules and Hamburg Rules.

Except in the situations where the carrier is exonerated from his liability for damage to goods or loss of goods as enumerated in Article 4, the carrier should compensate the cargo owner in case of damage of goods under his care.

The newly adopted Rotterdam Rules[24] mention even the delay in delivery of goods as situation (cause) leading to the carrier's liability.

Once damage to cargo (Loss of cargo) has occurred and the carrier held liable for the damage, the cargo owner will need to be compensated for the damage. In practice, the cargo underwriter

---

[24] Formally, the United Nations Convention on Contracts for the International Carriage of Goods Wholly or Partly by Sea, 2008, United Nations. The convention is not yet into force.

will pay the owner of the cargo. The cargo underwriter will then seek to recover his losses from the ship owner through subrogation.[25]

The ship owner will hand over the case of damage to his insurer who is a Protection and Indemnity Club known under the acronym "P&I Club."

One of the following international conventions shall determine the rules applicable to the damage:

> Hague Rules (1924),
> Hague-Visby Rules (1968) and
> Hamburg Rules (1978).[26]

In addition to the aforementioned maritime liability regimes, it would be relevant to mention the Rotterdam Rules which will be studied in relation to the risks generally covered in Protection and Indemnity Insurance.

In this paper the evaluation made in the maritime liability regimes is considered in their impact on marine insurance law & practice. It is assumed that no party involved in cargo claims can master

---

[25] The doctrine of subrogation states that the insurer may step into the shoes of the assured and enforce any claim, defence or set-off the assured possesses against any third party; the right arises only after the insurer has paid for the loss.

For further details, CHUAH (Jason), *Law of International Trade,* 3rd Edition, p.p. 448-452

[26] United Nations Convention on the Carriage of Goods by Sea, signed at Hamburg on March 31, 1978, and in force November 1, 1992

the legal complexity around the question dealing with the ship owner's liability with its impact in the insurance industry. Another challenge will be about the removal of the "vacuum areas" with regard to the insurance of loss of goods in seaborne trade. The analysis of marine cargo claims will illustrate some difficulties in this respect.

Therefore, the study is restricted to the following issues: compensation of damages, limitation of liability, notification of the damage to cargo (Loss of cargo), time limitation of action (effect of the Time-Bar) and the impact of the Rotterdam Rules on the marine insurance industry. Since the liability of the ship owner is covered by the P&I insurance policy, what role will be played by Cargo Insurance? Is Cargo Insurance needed in international trade?

SOME WORDS ON THE MARINE INSURANCE INDUSTRY

This section is part of the introduction for the only purpose of a better understanding of the subject.

Two aspects of insurance are considered here: third party liability for cargo claims offered in Protection and Indemnity marine insurance (P&I) and Cargo Insurance which is offered on the open market of insurance.

Cargo Insurance

The primary function of Cargo insurance is indemnification of claimants for proven loss of cargo or damage to goods.[27]

Cargo insurance helps protect the interest of the shipper and keep transport costs to a minimum. It has the advantage that it can be exactly tailored to meet the requirements of the Shipper with regard to the transport risks covered and the sums insured[28]

It is worth pointing out that in any export/import contract consideration should be given to the relationship between the insurance cover and the contract of sale.

It is argued that Cargo insurance, with respect to losses and damages for which carriers are responsible, is a banking service, in the sense that the cargo owner receives the funds promptly and remains in business with a minimal disruption of their business without using their own financial resources or credit. In contrast with the compensation mechanism in Cargo insurance, the present

---

[27] Hodges (Susan), *Cases and Materials on Marine Insurance Law*, Cavendish Publishing Limited (London. Sydney), p.p.: 1-2, 2002

[28] Vitiritti (Luis), 'Zurich Marine Risk Insight 2011', Insights in Marine Risk, June 2011, p. 2; McDOWELL (Carl E.), 'Containerization: Comments on Insurance and Liability',3 JML & C. 500 1971-1972 (HeinOnline), p.p. 509-512

system dealing with carriers' liability often does not provide a prompt compensation[29]

Thus, it can be concluded that Cargo insurance has a positive impact on the development of international trade.

PROTECTION AND INDEMNITY

P&I Insurance has traditionally been provided by mutual ship owners' associations called P&I Clubs which appeared in the 1850's. The Clubs are mutual insurance associations of which the ship owners are the members. They are owned by, and run for the benefit of their members. In contrast with commercial marine insurers who are answerable to their shareholders, the Clubs are run as non-profit-making businesses.[30] The Group Clubs compete between themselves. Nevertheless, they share between them liabilities in excess of 8 million US Dollars up to a maximum limit of around 7 billion Dollars through the structure of the International Group Agreement.

The International Group basically performs three key functions: the operation of the claims pooling and reinsurance programme, providing a forum for the exchange and consideration of views

---

[29]   Hodges (Susan),Op. Cit., p.p. 87-99; McDOWELL (Carl E.), Op. Cit., p.p. 509-510.

Please, note that The Institute Cargo Clauses (A),(B) and (C) are of particular relevance here.

[30]   IGP&I, Op. Cit., p. 3

on issues relating to ship owners' marine liabilities and insurance arrangements, and external representation.

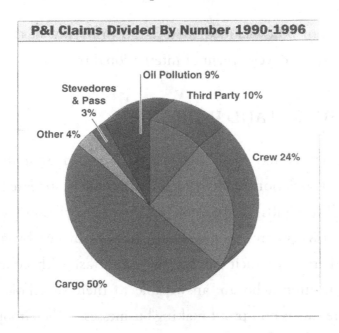

As to the general financial situation of P&I Clubs, the following trend has been observed for the last five years.[31]

Few clubs reported an underwriting surplus and the risks were both increasing and getting more complex due to new requirements in relation to the environmental protection and the need for more safety, including security in maritime transportation.

Facing the economic crisis in 2008, many clubs experienced financial difficulties with their reserves going significantly down: except the North of England, the Newcastle based Protection & Indemnity Club and Gard (in Norway), most of the clubs had to settle very high claims.

Moreover the entered tonnages (insured ships) were getting older and, as a result of that, many more damages occurred.

However, the prevailing situation in 2009 was characterized by a financial recovery that has been taking place in the Insurance industry until now. The International Group of P&I clubs has been pursuing general increases of between 12.5% and 29% while members were struggling with the sharp decrease in the freight and financial market.

As an illustration of the described situation, six of the mutuals-American, London, Steamship, Swedish, UK and West of England - made unbudgeted call as pressure in the financial market reduced their investments portfolios (reserves).

---

[31] Frank (Jerry), 'North of England, a winner in annual renewals', Lloydslist.com, 20-02-2009

The last two years have been marked by recovery and stability in the marine insurance industry.

A new trend with particular relevance to P&I insurance is both the advent of "Independent P&I Clubs" in the newly industrialized countries (China, India and South Korea) and the presence of "the commercial insurers of P&I liabilities." These new comers are offering worthy P&I insurance covers.

The presence of these independent actors and high competition coming from P&I facilities and other insurance schemes more suitable to the new trends in the transportation industry provided on the commercial market have increased the level of competition in P&I insurance which is moving from a situation of oligopoly to a fairly open market.

LEGAL IMPACTS

1. Compensation of damage: calculation and type of damage

The three maritime carrier liability conventions were adopted at different periods of time and were conceived differently with regard to the concept of damages and the methods of damage evaluation.[32]

---

[32]   Bauer (R. Glenn), 'Conflicting Liability Regimes: Hague-Visby v. Hamburg Rules – A Case by Case Analysis', p. 53, HeinOnline- 24 JML & Com.53 1993; Hoeks (Marian), *Multimodal Transport Law: The law applicable to the multimodal contract for the carriage of goods*, p.p. 81-85, 2010

The Hague-Visby Rules contain some instructions on the calculation of damages. In spite of these instructions, it could be difficult to get their real meaning, for example Article 5 (b) of the Hague-Visby Rules. What is the meaning of the sentence: "...total amount recoverable shall be calculated by reference to the value of such goods at the place and time at which the goods are discharged." The Report of the President of the Commission suggests that the new rule should be understood in the context of harmonisation of the calculation of damages.

In contrast with the Report, English case-law suggests that Article 5 (b) simply lays down a prima facie measure of damages: the damages have to be assessed according to appropriate circumstances and can include the recovery of additional damages.

The question of additional damages leads to controversy.[33]

However, most jurisdictions and good practice in insurance generally assess the real circumstances and if, found appropriate, should be included in the recoverable amount.[34]

In the Hamburg Rules, the matter is left to the applicable law. The main issue here will be how to find this applicable law knowing the difficult context of the carriage of goods regimes.

---

[33]   von Ziegler (Alexander), 'Compensation of Damage', Paper presented at the conference on the Rotterdam Rules, Rotterdam, 24 september 2009 (Concept of financial impact of the loss

[34]   Standard rule of practice in P&I insurance (This can be included in a policy or the Club Standard rules)

What about the Rotterdam Rules?

RR, Article 22 (3): The provisions can be understood as clarification that no further weight or items are considered unless agreed.

Research on relevant cases shows that cases are decided in the spirit of including all potential and possible claims in the restrictive rules of the carriage conventions. [35]

2. Limitation of liability

Limitation of liability is inherent to the international seaborne trade. It was conceived in order to encourage owners to engage their resources (Crew members and sea-going ships) in carriage of goods in spite of the many hazards of the sea and protect them from unlimited exposure to liability.

There is a variation of limits in the maritime liability regimes with package limit and weight limit. With regard to package limit, the amount per convention is as follows 666.67 Hague-Visby Rules, 835 in Hamburg Rules and 875 in the Rotterdam Rules.[36]

---

[35] Tetley (W), Op. Cit; and Hazelwood (Steven J.), *P&I CLUB, Law and Practice, 2010*

[36] Von Ziegler (Alexander), 'Compensation of Damage', Paper presented at the conference on the Rotterdam Rules, Rotterdam, 24 September 2009

Taking into account the weight limit, the distribution is as the following: 2 in Hague/Hague-Visby Rules, 2.5 in Hamburg Rules and 3 in the Rotterdam Rules.[37]

A legal issue of interest is the discussion as to whether a container is a package or is it the goods contained therein which are packages.

Courts interpreting the Hague Rules have had difficulty in assessing whether a container should be a package for the purpose of limitation of liability. It generally depends on the descriptions in the bill of lading. In case the bill of lading describes the container as one package without listing its contents, the container itself is the package. However, in the case where the bill of lading includes a reference to the contents inside the container, the contents determine the number of packages. According to Article 2 (C) of Hague-Visby Rules, the container will be a package or unit if the number of packages or units inside are not enumerated in the bill of lading.[38]

The Rotterdam Rules address this issue at Article 59 and the principle of solution is the following: "The packages or shipping units enumerated in the contract particulars as packed in containers or pallet are deemed packages or shipping units.

Another concern with respect to the container in the awake of damages due to defective lashing systems leading to very costly

---

[37]   Von Ziegler (Alexander), Op. Cit.

[38]   Tetley (William), Op. Cit, p.p.: 1543-1545 (with report of relevant cases in different jurisdictions)

maritime disasters is to assess whether a slot charterer is entitled to the same limitation of liability as the ship owner. In Metvale Ltd v Monsanto International and others, Mr Justice Teare decided that slot charterers were entitled to limit liability under the convention[39]

As to the validity of limitation clauses of liability, these clauses are not accepted in case the damage or loss has occurred by negligence from the carrier.[40]

3. Notice of damage

Notification of damage to the carrier is a prerequisite to any claim from a breach of contract in carriage of goods. Failure from the consignee or the shipper to give timely notice of damage to the carrier may result in the loss of the rights for compensation.

The rules dealing with the notice of damage vary according to the different conventions. It is required from the shipper or the consignee to give a notice of loss or damage in writing to the carrier or his agent at the port of discharge before or at the time of the removal of the goods into the custody of the person entitled to delivery according to the contract of carriage in accordance with both the Hague and the Hague-Visby Rules. In case the loss or damage is not apparent, the notice should be given within three

---

[39] Please, read also Hjalmarsson (Johanna), 'Shipping & Trade Law', 200 9, (2), 3-4

[40] Case 515 US 528, 115 S. Ct, 23 22, 132L. Ed. 2d 462 – Supreme Court, 1995- Google Scholar: Loss of right to limitation of liability for ship owner' s negligence

days of the delivery. There is no need for notice of damage in the case where the state of the goods has been subject of joint survey or inspection at the time of their reception.[41]

In the Hamburg Rules, the notice of loss or damage is specified at article 19. The notice should be given in writing by the consignee to the carrier not later than the working day after the day when the goods were handed over to the consignee.

The Rotterdam Rules are more explicit about damage notification: it should be given within 21 consecutive days after delivery of the delayed goods (Art. 23, (4), RR). The content of the notice in cases of delay should consist of:

*information on delay (as damage) and
*the fact that damage follows from delay.

Besides the notification of damage, there is a delay within which a claim may be made against the carrier.

4. Limitation of Action – Effect of the Time Bar

In application of the Hague and Hague-Visby Rules, the time-bar limit related to claims against the maritime carrier with regard to liability for loss of good or damage to goods is one year (Article III). In the case of any actual or apprehended loss or damage, the carrier and the consignee (shipper) should give all reasonable facilities to each other for inspecting and tallying the goods.[42]

---

41 Article III (9) of Hague-Visby Rules

42 Article III (5) of Hague-Visby Rules

In the Hamburg Rules the delay for claims has been fixed to two years (Art. 20).

In addition to this principle, the person against whom a claim is made may at any time during the running of the limitation period extend that period by a declaration in writing to the claimant. This period can be further extended by another declaration or declarations.[43]

These provisions are very often used by both parties once a marine cargo claim has been introduced. In that regard, careful attention should be given to all the documents exchanged between parties for extension of delay.[44]

Cases

It was held in "Aries Tanker Corp. v Total Transport Ltd" that the carrier was discharged from all liability in application of the rules on limitation of Action in Hague-Visby Rules.[45]

This principle was confirmed later in the case "Bua International Ltd v Hai Hing Shipping Co Ltd.[46]

The principle that the limitation period applies regardless of whether the claim is founded in tort or contract is found in the

---

[43] Article III (5) of Hague-Visby Rules

[44] As Standard Rule of Practice in P&I Insurance

[45] The Aries (1977) 1, Lloyd's Rep. 334, by the House of Lords

[46] The Hai Hing (2000) 1, Lloyd's Rep. 300

case Salmond and Spraggon (Australia) Pty Ltd v Port Jackson Stevedoring.[47]

5. Implications of the Rotterdam Rules in P&I insurance and Cargo Insurance

The newly adopted Convention on Contracts for the International Carriage of Goods Wholly or Partly by Sea which is also called "Maritime Plus" instrument is intended to replace the three existing conventions relating to the ship owner liability.

In spite of the desirability of the rules established in the new convention, there is no unanimity as to the future effectiveness of the rules.

P&I Clubs are of the general opinion that the rules may add some extra costs in marine insurance.[48]

However, they believe that a single liability regime would speed claims payments and reduce claims costs in the long term: a single standard would reduce uncertainty and enable claims to be settled faster.

The rules are largely debated by practitioners, legal experts and scholars with deep insight in issues relating to the harmonization

---

[47] The New York Star, (1980) 2, Lloyd's Rep., 317

[48] This has always been the position of the P&I Clubs during the process of elaboration of the Rotterdam Rules

of the rules of carriage of goods and marine insurance law & Practice.[49]

Before engaging in the discussion and arguments pertaining to the need and the viability of the adopted convention, it is advisable to identify the specific provisions in the Rotterdam Rules relevant to global challenges to be addressed in the development of the Insurance industry. The rules are designed to regulate both the international maritime carriage of goods and the international multimodal carriage of goods including a sea leg.

The term a "maritime plus" instrument refers to the new technology in the Convention, changes to some traditional terminology and familiar terms (For example, "bills of lading" and "sea waybills" are now included in the general term "Transport Document" in the Rotterdam Rules.

What can be the consequence of a possible ratification of the Convention in cargo claims?

It is argued that ratification will significantly increase the liability of ship owners and maritime carriers in respect of the carriage of goods. In particular, the long established exception of error in the navigation or management of the ship (the nautical fault exception) would be excluded. The obligation to exercise due

---

[49]   D. Rhidian (Thomas), 'And then there were the Rotterdam Rules', (2008) 14 JIML, at pp. 189-190; Eftestol-Wilhelmsson (Ellen), 'The Rotterdam Rules in a European Context ', (2010), 16 JML

diligence and to provide a seaworthy ship has been extended to the duration of the voyage instead of being restricted to before and at the commencement of the voyage as specified in the Hague/Hague-Visby.[50]

The limits of liability per package or unit of weight have been significantly increased beyond Hague-Visby and Hamburg Rules limits.[51]

The liability of the owners and other maritime carriers for the negligence of maritime performing parties would be regulated by the Rules.[52]

At present, it is not possible to evaluate the economic impact of these increased liabilities, but it is obvious to mention that ship owners and their P&I insurers would see a significant increase in the cost of cargo claims in case the Rotterdam Rules are adopted globally.

The convention contains a number of new and effective features. The scope of application of the convention would extend to

---

[50] Rotterdam Rules as seen by the P&I Club West of England, Luxembourg, Review, 2012

[51] This can be seen through comparison of art. 4 (5) Of Hague Rules, Article 4 (5a) of Hague-Visby Rules, Article 6 (1a) of Hamburg Rules, and Article 59 (1) of Rotterdam Rules

[52] It should be mentioned here that the NVOOC are considered as maritime carriers

door-to-door carriage and tackle-to-tackle/ port-to-port carriage and many of the beneficial aspects of existing conventions and regimes are retained. In particular, it retains the existing concept of network liability, whereby liability and the applicable limits of liability for loss of and damage to the goods occurring before or after the sea-leg will be a matter of any unimodal international instrument compulsorily applicable to the relevant mode of transport where the loss or damage occurs.

The Rotterdam Rules retain the concept of fault-based liability found in the Hague/Hague-Visby Rules, but the standards and burdens of proof overall are more onerous for the carrier. The Convention has provisions for electronic commerce. It also allows more freedom of contract in the liner trade by introducing the concept of "Volume Contract" in carriage of goods.

Similarly to the existing conventions, the Rotterdam Rules would apply to transport documents, such as bills of lading and sea waybills, issued both in liner and in non-liner trades.

The Rotterdam Rules address the issues relating to jurisdiction and arbitration: the relevant provisions are essentially based on the restrictive approach of the Hamburg Rules. Under the Rules cargo owners can effectively choose from a number of jurisdictions the court where they can sue the carrier.

As to the exclusive jurisdiction agreements contained often in contracts of carriage, they do not have primacy. However, these provisions are subject to an "opt in" by States.[53]

The ability of cargo owners to choose from a number of jurisdictions may lead to greater uncertainty for carriers and insurers and high legal costs due to the fact that courts of countries unfamiliar with such issues are asked to decide test cases arising under the Convention.

As it appears, the Rotterdam Rules seem to be an ambitious project trying to codify almost all aspects of carriage of goods by sea. There seems little doubt that the status quo of existing regimes will not remain in case of no ratification of the Rules: the EU and the United States would enact their own domestic legislation.

While advising the International Group of P&I Clubs (IG), the International Chamber of Shipping (ICS) and the European Community Ship Owners' Associations (ECSA), the European Commission indicated that they remain open minded with regard to the question of the desirability of a European multimodal convention. The European Commission has publicly criticized the Rotterdam Rules. The reason for that is that the rules do not conform to European multi-modal expectations.[54]

---

[53] The Rotterdam Rules as seen by the P&I Club West of England, Luxembourg, Review, 2012

[54] The Rotterdam Rules as seen by the P&I Club West of England, Luxembourg, Review, 2012

Generally speaking, it is desirable that the Rotterdam Rules enter into force. The number of ratifications required for that has been set at 20.

If the Rotterdam Rules can have any real impact, it will need to be adopted by a large number of States. Therefore, it can be mentioned here as example that approximately 90 States ratified the Hague the Hague/Hague-Visby Rules but only 34 the Hamburg Rules. Some of the 34 signatories which ratify the Hamburg Rules were not major trading nations.

Standard for Protection and Indemnity Cover

Club cover for liability in respect of cargo is based on the relevant contract of carriage which is subject to terms no less favourable than the Hague/Hague-Visby Rules.

The Club's Cargo standard Rule generally provides as follows:

"…there is no cover in respect of liabilities which would not have been payable by the Member if the contract of carriage had incorporated the Hague Rules, the Hague-Visby Rules, or similar rights, immunities and limitations in favour of the Carrier. There is no cover in respect of liabilities arising under the Hamburg Rules unless the Hamburg Rules are compulsorily applicable to the contract of carriage by operation of law."[55]

---

[55] Standard Rule in P&I Marine Insurance

Club cargo cover is based on the liability regime set out in the Hague/Hague-Visby Rules mainly because it is recognised that this has been the accepted standard for a long time, standard according to which the majority of ship owners and cargo owners contract commercially, when contractual terms are not imposed by law. Besides, the Hague/Hague-Visby Rules have been ratified or (as in the case of the USA) adopted as ground for domestic legislation by a substantial number of States and mainly, by major trading states. It is worth remembering that when the Hamburg Rules came into force, the clubs did not substitute them for the Hague/Hague-Visby Rules. The reasons for that are that those states which had ratified the convention are not major trading nations and the industry was not interested in the Hamburg Rules.[56]

The legal issue which will arise is to what extent P&I cover should be available for liabilities incurred by Members in the context of the Rotterdam Rules being ratified?

The provisions of the new Convention would normally become compulsorily applicable by law in the relevant state where that state either ratifies the Convention and the Convention has entered into force or enacts its provisions by other means into its domestic law (for example US COGSA). Hence, the Club would indemnify a member in respect of cargo liabilities arising compulsorily by law under the new Convention.

---

[56] The main trading countries still prefer the maritime carrier liability regime established by the Hague-Visby Rules

However, if the Member had voluntarily agreed to the incorporation of the Rotterdam provisions into the contract of carriage, the position of the Club would be different: the Club would not cover liabilities arising under the Rotterdam Rules in excess of those which would be covered according to the Hague/Hague-Visby rules.

At present, there is uncertainty about the future of the Rotterdam Rules?

Would the major maritime trading countries ratify the new Convention?

If they do ratify these rules, would they incorporate all the rules into their domestic law?

Many scholars have been discussing the aforementioned questions; it seems advisable to mention some studies of relevance[57]

In addition to that, it remains to be seen to what extent carriers will be willing to adopt the Rotterdam Rules as the standard for their carriage terms, and, with particular reference to the marine insurance industry, there may be differing opinions from those engaged in liner shipping as opposed to other sectors.

---

[57] Eftestol-Wilhelmsson (Ellen), 'The Rotterdam Rules in a European Multimodal Context' (2010), 16 JML; D. Rhidian (Thomas), 'And then there were the Rotterdam Rules', (2008) 14 JIML, pp. 189-190

Another uncertainty factor is the position of the International P&I Group. It is not presently known whether and, if so, when the Group Clubs will consider it appropriate to replace the Hague/Hague-Visby Rules with the Rotterdam Rules as the standard for P&I cover after ratification of the Rules. The views of the Clubs' membership as a whole will be a determinant factor in the process of the voluntary incorporation of the Rotterdam Rules in the carriage contract. It is highly suggested that the Clubs keep the matter under regular review.[58]

As to the current situation, a Member who is in need to be covered for liabilities which he has voluntarily accepted and agreed beyond the normal standard of the Hague/Hague-Visby Rules can ask his Club to arrange cover for the additional liabilities on special terms by agreement (Here is an affirmation of the principle of freedom of contract tailored to the need of the Member). [59]

---

[58]   IGP&I, Op. Cit., pp:4-11

[59]   Standard Rule in P&I Marine Insurance

CHAPTER VI

BILL OF LADING: ESSENCE AND EVOLUTION THROUGH ELECTRONIC

DOCUMENTS

In the preceding chapters the main issues were actually about codification, standardization and harmonization of the rules pertaining to the bill of lading in the International Seaborne Trade. The Hague Rules are officially referred to as "International Convention for Unification of Certain Rules of Law in Relation to Bills of Lading" (Brussels Convention on the Bill of Lading).

The HVR are a protocol to the Brussels Convention (1924).

Moving forward, the Hamburg Rules introduce the Sea Waybill as negotiable document in shipping whereas the Rotterdam Rules recognize the use of electronic documents together with traditional bills of lading in International Seaborne Trade.

In this essay the Bill of Lading is mainly considered in the context of International Maritime Law. That means proof of the maritime carriage contract.

Since, there seems to be an abundant legal literature about the Bill of Lading this work will focus more on the importance of this maritime transport document in Marine Insurance and Practice.

DEFINITION OF THE BILL OF LADING

It is a document signed by a carrier (a transporter of goods) or the carrier's representative and issued to a consignor (the shipper of goods) that evidences the receipt of goods for shipment to a specified destination and person.[60]

Carriers using all the modes of transportation issue bills of lading when they undertake the transportation of cargo.

A bill of lading is, in addition to a receipt for the delivery of goods, a contract for their carriage and a document of title to them. Its terms describe the freight for identification purposes; state the name of the consignor and the provisions of the contract for shipment; and direct the cargo to be delivered to the order or assigns of a particular person, the consignee, at a designated location.

Here are some basic types of bills of lading.

---

[60]   Dictionary/Thesaurus

A straight bill of lading is the one in which the goods are consigned to a designated party.

Besides that, an order bill of lading presents the particularity of using express words to make the bill negotiable; it states that delivery is to be made to the further order of the consignee.

In addition, there is a Bearer Bill of Lading which states that the delivery shall be made to whosoever holds the bill.

The interest of this distinction lies in determining whether a bill of lading is negotiable (capable of transferring) title to the goods covered under it by its delivery or endorsement. If its terms provide that the freight is to be delivered to the bearer (or possessor) of the bill, to the order of a named party or, as recognized in overseas trade, to a named person or assigned, a bill, as a document of title, is negotiable.

In The United States of America, State laws, which often include provisions from "the Uniform Commercial Code", regulate the duties and liabilities imposed by bills of lading covering goods shipped within state boundaries.

Federal Laws, embodied in the **Interstate Commerce Act** (49 U.S.C, 1976 Ed, §1 et seq.) apply to bills of lading covering goods carried (moving) in interstate commerce.[61]

---

[61]   West's Encyclopedia of American Law, Edition 2, Copyright 2008, The Gale Group. Inc. All rights reserved

OTHER DEFINITIONS OF THE BILLS OF LADING

"The bill of lading is a receipt obtained by the shipper of goods, from the Carrier (trucking company, railroad, ship or air freighter) for shipment to a particular buyer. It is a contract protecting the shipper by guaranteeing payment and satisfies the Carrier that the recipient has proof of the right to the goods."

The bill of lading is then sent to the buyer by the shipper upon payment for the goods, and is thus proof that the recipient is entitled to the goods when received.

Thus, no delivery of goods without bill of lading.[62]

"The bill of lading is a memorandum or acknowledgment in writing, signed by the captain or master of a ship or other vessel, that he has received in good order, on board of his ship or vessel, therein named, at the place therein mentioned, certain goods therein specified, which he promises to deliver in like good order (the dangers of the seas excepted) at the place therein appointed for the delivery of the same, to the consignee therein named or assigned, he or they paying freight for the same."[63]

---

[62] N. Hill and Kathleen T. Hill, Copyright © 1981-2005; see also INVOICE in Burton's Legal Thesaurus, 4 E, COPYRIGHT © 2007 by William C. Burton

[63] 1 T.R. 745. Bac. Abr. Merchant L Com. Dig. Merchant E 8.b; Abbot on Ship. 216 1 Marsh. On Ins. 407; Code de Com. Art 281

"The bill of lading is the written evidence of a contract for the carriage and delivery of goods sent by sea for a certain freight." [64]

The bill of lading ought to contain the name of the consignor, the name of the consignee, the name of the master of the vessel, the name of the vessel, the place of departure and destination, the rate of the freight; and in the margin, the marks and the number of the things. It is usually made in three originals or parts. One of them is commonly sent to the consignee on board with the goods; another is sent to him by mail or some other conveyance; and the third is retained by the merchant or shipper. The master should also take care to keep another part for his own use.

Moreover, the document is assignable. And the assignee is entitled to the goods, subject to the shipper's right in some cases of stoppage in transit.

"The bill of lading is a document whereby a transport company acknowledges the receipt of the goods and serves as title for the purpose of transportation."[65]

It could be considered as a written receipt issued by a carrier, a transport company, stating that it has taken possession and received an item of property and usually also confirming the details of delivery (such as method, time, place or to whom) and serves as the carrier's title for transportation purposes.

---

[64]   Per Lord Loughborougb, 1 H. Bl. 359

[65]   References: Aman v. Dover & Southbound R. Co. 179 NC 310 (1920)

In Mills & Co, English judge Selbourne wrote:

"...The primary purpose of a bill of lading, although by mercantile law and usage it is a symbol of the right of property in the goods, is to express the terms of the contract between the shipper and the ship owner."

In Aman, the Supreme Court of North Carolina defined the word (bill of lading) and added the statement of the law that a bill of lading is not a necessary thing for the carrier to be liable for the safe delivery of the item of property.

"... An instrument issued by the carrier to the consignor, consisting of a receipt for the goods and an agreement to carry them from the place of shipment to the place of destination, is a bill of destination, is a bill of lading."[66]

"Every bill of lading in the hands of a consignee or endorsee for valuable consideration, representing goods to have been shipped on board a vessel or train, is conclusive evidence of the shipment as against the master or other person signing the bill of lading, notwithstanding that the goods or some part thereof may not have been shipped, unless the holder of the bill of lading has actual notice, at the time of receiving it, that the goods had not in fact been laden on board, or unless the bill of lading has a stipulation to the contrary, but the master or other person so signing may exonerate himself in respect of such misrepresentation by showing that it was caused without any default on his part, and

---

[66]   See ref.65, Idem

wholly by the fault of the shipper or of the holder, or of some person under whom the holder claims.[67]

EVOLUTION OF THE BILL OF LADING AND ELECTRONIC DOCUMENTS

As to the evolution of the bill of lading, the following trend has been observed: documentation is basically an area of detail which depends on legislation and international commercial procedure both of which are subject to change.[68]

The main issues associated with documentation are related to speed and expenses.

Recently, a considerable progress has been made in order to solve the problem: standardization of the required documents has been achieved and both the cost and the time needed for processing of these documents have been considerably reduced.

In October 1960 the Committee on the Development of Trade of the Economic Commission for Europe decided to set up a working group to consider the possible reduction, simplification and standardization of trade documents.

---

[67]   Ref. 65, Idem

[68]   Alderton (Patrick), *Sea Transport: Operation and Economics (Documentation)*, Chap XIV, Surrey, UK, Thomas Reed Publications, 1995

By the mid-sixties many organisations were initiating ideas related to aligned which involve using documents of a standard size and a standard format.

In 1963 the ICS (International Chamber of Shipping) through the Facilitation Committee produced its first report on "Standard Format for Bills of Lading."

Nevertheless, modern Bills of Lading are based on the ICS 1972 Standard Bill of Lading.

In 1981 an EEC Documentation study showed that for a container moving from UK to USA 128 different documents could be involved of which one third were official and two thirds were operational.

SITPRO (SIMPLIFICATION OF TRADE PROCEDURES)

The SITPRO Board was set up in 1970 to guide, stimulate and assist the rationalisation of international trade procedures, the documentation and information flows related to them, namely in view of the widening use of computers and telecommunications links. SITPRO members are shippers, carriers, forwarders, bankers, insurers and government representatives. Its working groups call on the specialist expertise of many commercial and official interests. Funds are provided by the Department of Trade through an annual grant.

However, the Board enjoys independence within its agreed terms of reference.

In 1975 the Special Program on Trade Facilitation in UNCTAD was set up and was mainly instrumental in order to get worldwide acceptance of the 1972 ICS Standard Bill of Lading.

Currently, there is a growing use of computer links with the customs for clearing documents.

The International Maritime Bureau (IMB) was set up to counter international maritime fraud.

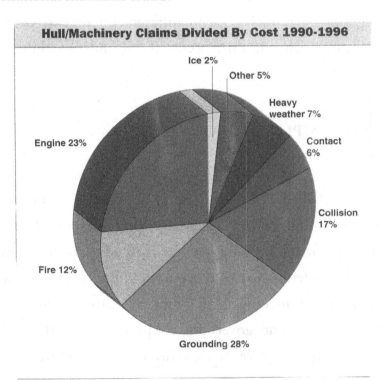

CONCLUSION

The standard Club Cargo Cover is currently based on the liability regime set out in the Hague/Hague-Visby Rules: these rules have been recognised for a long time ago as the accepted standard according to which the majority of maritime carriers and cargo owners contract commercially when contractual terms are not imposed by law.

Another factor in justification of the recognition of the Hague/Hague-Visby regimes lies in the ratification or adoption of these rules as the basis for domestic legislation by a significant number of states, including the major trading countries.

Insofar, the marine insurance industry has been reluctant to adopt the Hamburg Rules as standard regime in Third Party liability and in Cargo Insurance.

The majority of developing countries ratified the Hamburg Rules and are positive to their application. An interesting aspect to mention is that the application of the Hamburg Rules has enhanced the work associated with the management of the cargo through the creation of Central Freight Bureaux which could be an effective tool for the promotion of the International Trade and economic investments for the maritime infrastructure in those countries.

Another point which is worth noting is the position of the Nordic countries expressed in their maritime legislation: Denmark, Finland, Norway and Sweden apply some rules from the Hamburg Rules incorporated in their national law only to maritime issues in carriages involving countries which ratified the Hamburg Rules. It is made clear that they are still parties to the Hague-Visby Rules. Hence, questions of conflicts of law involving most countries with major parts in International Trade (USA, UK, Germany, Canada, etc…) would be settled according to the Hague-Visby Rules (Cfr The Swedish Maritime Code).

As to the Rotterdam Rules, they are not yet into force; there is uncertainty on whether, if ratified, they can be adopted as standard in the marine insurance industry. Nevertheless, the rules can contribute to a substantial reduction in claims in virtue of harmonisation of the rules relating to the carriage of goods by sea in a single regime in the long run.

Currently, a maritime carrier or a cargo owner can afford the needed insurance cover by agreement with the insurance provider.

The RR certainly mark a legal innovation in the rules governing the international transport sector. There is no doubt on this point.

What is at stakes is whether or not the RR will be ratified. Furthermore, once ratified will the major maritime powers accept them?

The desirability of creating a new international legal regime to regulate the carriage of goods in international trade is recognized all over the world. As explained in the previous chapters, the three existing international maritime regimes (HR, HVR and Hamburg Rules) are widely considered as outdated. Furthermore, there is a need for international uniformity so that major trading nations could apply the same law to improve administration of the law and reduce transportation costs. The RR concluded by UNCITRAL and adopted by the United Nations General Assembly in 2009, are proposed as a new international transport legal regime for the twenty-first century that considers containerization of cargo, multimodalism and electronic transportation documents.

It is argued that while on balance the Rotterdam Rules represent a marginal improvement over the patchwork of existing international transport regimes, they also seem to present many difficulties. The RR are far from ideal, however they offer a "next best" solution that would contribute to the improvement of the multimodal international carriage of goods if adopted globally to replace the three existing maritime legal regimes.

Then, what could be the criteria by which to judge the Convention (RR)?

Five criteria could be used in this assessment. They are presented as follows.

Firstly, it would be necessary to know whether the Rules enhance certainty and predictability in International Transport Law or not.

Secondly, do they offer a higher degree of satisfaction as far as the commercial needs are concerned?

The interest of the question lies in the fact that such rules should address innovative issues in International Trade, namely: the electronic bills of lading and multimodalism.

The third criterion is associated with the need for minimizing transactional litigation costs (the questions of conflicts of Law and Jurisdiction).

Further, the question of enhancing distributive justice among the stake holders in International transportation would also show the importance of the RR.

The fifth criterion, the last one, is whether the Convention could promote International uniformity in the International Carriage of Goods or not. This only time will tell.

Finally, the role played by the maritime surveyor in Protection and Indemnity Marine Insurance and Practice needs to be mentioned. An efficient International P&I club would work closely with surveyors who are specialists in assessing maritime damages and operating all over the world. As soon as a damage occurs, they should be contacted for a reliable account associated with maritime casualties. Actually, upon a notice of damage from the ship owner to the club, the insurance provider should appoint a maritime surveyor in order to assess the cargo damage (to determine the reasons behind the damage).

BIBLIOGRAPHY

1) Textbooks

Alderton (Patrick), *Sea Transport: Operation and Economics,* Surrey, UK, Thomas Reed Publications, 1995

Chua (Jason), *Law of International Trade,* Third Edition, Thomson Sweet & Maxwell, London 2005

Hazelwood (Steven J.)., *P&I Club: Law and Practice, 2010*

Hodges (Susan), *Cases and Materials on Marine Insurance Law,* Cavendish, London Sydney, 2002

Hoeks (Marian), *Multimodal Transport Law: The law applicable to the multimodal contract for the carriage of goods, 2010*

Tetley (William), *Marine Cargo Claims,* Fourth Edition, Volume 1, Thomson*Carswell, 2008

Tetley (William), *Marine Cargo Claims,* Fourth Edition, Volume 2, Thomson*Carswell, 2008

2) Articles

Bauer (R. Glenn), 'Conflicting Liability Regimes: Hague – Visby v. Hamburg Rules – A Case by Case Analysis' 24 JML & Comm., 531

Crowley (ME)-Tul. L. Rev., 2005, 'The Uniqueness of Admiralty and Maritime Law: The Limited Scope of the Cargo Liability Regime covering Carriage of Goods, Multimodal Transport', 2005

D. Rhidian (Thomas), 'And then there were the Rotterdam Rules', (2008), JIML

Derrington (Sarah), 'Marine Insurance Act 1906 – Introduction', web site www.findlaw.com.au (articles) 12/04/2012

Eftestol-Wilhelmsson (Ellen), 'The Rotterdam Rules in a European Context', (2010), 16 JIML

IGP&I (International Group of P&I Clubs), Annual Review, 2011-2012, London

McDowell (Carl E.), 'Containerization: Comment on Insurance and Liability', 3. JML & C 500 1971 – 1972

Sampani (Constantina), Aberystwyth University & Stephen Wrench, Lloyd's Maritime Academy, Informa plc, 'Reflexions on Legal Issues in a Diverse Maritime World', Cambrian Law Review, Volume 44, 2013

Vitiritti (Luis), ' Zurich Marine Risks', Insight 2011, June 2011, Zurich

Rosaeg (E), 'The Applicability of Conventions for the Carriage of Goods and for Multimodal Transport' in Lloyd's Maritime and Commercial Law Quarterly, 2002

3) Conference papers

Tshilomb (J), 'Multimodal Transport in relation to Ship Owner's Liability with regard to Cargo Claims', Seminar, Edinburgh Research & Innovation, The University of Edinburgh, May 2012

Von Ziegler (Alexander), 'Compensation of Damage', Paper presented at the Conference on the Rotterdam Rules, Rotterdam, the 24th of September 2009

Printed in the United States
By Bookmasters